Daines Barrington

An Essay on the Periodical Appearing and Disappearing of Certain Birds, at Different Times of the Year...

Daines Barrington

An Essay on the Periodical Appearing and Disappearing of Certain Birds, at Different Times of the Year...

ISBN/EAN: 9783337196745

Printed in Europe, USA, Canada, Australia, Japan

Cover: Foto ©berggeist007 / pixelio.de

More available books at **www.hansebooks.com**

Received November 29, 1771.

XX. *An Essay on the periodical Appearing and Disappearing of certain Birds, at different Times of the Year. In a Letter from the Honourable* Daines Barrington, *Vice-Pref. R. S. to* William Watson, *M. D. F. R. S.*

DEAR SIR,

Read April 2, 9, 30, and May 14, 1772. AS I know, from fome converfation we have had on this head, that you confider the migration of birds as a very interefting point in natural hiftory, I fend you the following reflections on this fubject as they have occurred to me upon looking into moft of the ornithologifts who have written on this queftion.

It will be firft neceffary in the prefent, as in all other difputes, to define the terms on which the controverfy arifes. I therefore premife that I mean by the word Migration, a periodical paffage by a whole fpecies of birds acrofs a confiderable extent of fea.

I do not mean therefore to deny that a bird, or birds, may poffibly fly now and then from Dover to

Calais, from Gibraltar to Tangier, or any other fuch narrow ftrait, as the oppofite coafts are clearly within the bird's ken, and the paffage is no more adventurous than acrofs a large frefh water lake.

I as little mean to deny that there may be a periodical flitting of certain birds from one part of a continent to another : the Royfton Crow, and Rock Ouzel, furnifh inftances of fuch a regular migration.

What I mean chiefly to contend therefore is, that it feems to be highly improbable, birds fhould, at certain feafons, traverfe large tracts of fea, or rather ocean, without leaving any of the fame fpecies behind, but the fick or wounded.

As this litigated point can only receive a fatisfactory decifion from very accurate obfervations, all preceding naturalifts, from Ariftotle to Ray, have fpoken with much doubt concerning it.

Soon after the appearance of Monf. Adanfon's voyage to Senegal, however, Mr. Collinfon firft, in the Philofophical Tranfactions *, and after him the moft eminent ornithologifts of Europe, feem to have confidered this traveller's having caught four European Swallows on the 6th of October, not far from the African coaft, as a decifive proof, that the common fwallows, when they difappear in Europe, make for Africa during the winter, and return again to us in the fpring.

It is therefore highly incumbent upon me, who profefs that I am by no means fatisfied with the account, given by Monf. Adanfon of thefe European

* Part II. 1760, p. 459, & feq.

fwallows,

fwallows, to enter into a very minute difcuffion of what may, or may not, be inferred from his obfervation according to his own narrative.

I fhall firft however confider the general arguments, from which it is fuppofed that birds of paffage periodically traverfe oceans, which indeed may be almoft reduced to this fingle one, *viz.* we fee certain birds in particular feafons, and afterwards we fee them not; from which data it is at once inferred, that the caufe of their difappearance is, that they have croffed large tracts of fea.

The obvious anfwer to this is, that no well-attefted inftances can be produced of fuch a migration, as I fhall endeavour to fhew hereafter; but befides this convincing negative proof, there are not others wanting.

Thofe who fend birds periodically acrofs the fea, being preffed with the very obvious anfwer I have before fuggefted, have recourfe to two fuppofitions, by which they would account for their not being obferved by feamen during their paffage.

The firft is, that they rife fo high in the air that they become invifible *; but unfortunately the rifing to this extraordinary height, or the falling from it, is equally deftitute of any ocular proof, as the birds being feen during their paffage.

I have indeed converfed with fome people, who conceive they have loft fight of birds by their perpendicular flight; I muft own, however, that I have

---

* It is well known that fome ornithologifts have even fuppofed that they leave our atmofphere for that of the Moon. See Harl. Mifc. Vol. II. p. 561.

always

always fuppofed them to be fhort-fighted, as I never loft the fight of a bird myfelf, but from its horizontal diftance, and I doubt much whether any bird was ever feen to rife to a greater height than perhaps twice that of St. Paul's crofs *.

There feems to be but one method indeed, by which the height of a bird in the air may be efti-mated; which is, by comparing its apparent fize with its known one, when very near us; and it need not be faid that method of calculating muft depend entirely upon the fight of the obferver, who, if he happens not to fee objects well at a diftance, will very foon fuppofe the bird to be loft in the clouds.

There is alfo another objection to the hypothefis of birds paffing feas at fuch an extraordinary height, arifing from the known rarefaction of the air, which may poffibly be inconvenient for refpiration, as well as flight; and if this was not really the cafe, one fhould fuppofe that birds would frequently rife to fuch uncommon elevations, when they had no oc-cafion to traverfe oceans.

* Wild geefe fly at the greateft height of any bird I ever happened to attend to; and from comparing them with rooks, which I have frequently looked at, when perched on the crofs of St. Paul's, I cannot think that a wild-goofe was ever diminifhed, to my fight at leaft, more than he would be at twice the height of St. Paul's, or perhaps 300 yards. Mr. Hunter, F. R. S. in-forms me, that the bird which hath appeared to him as the higheft flier, is a fmall eagle on the confines of Spain and Portugal, which frequents high rocks. Mr. Hunter hath firft feen this fpecies of eagle from the bottom of a mountain, and followed it to the top, when the bird hath rifen fo high as to appear lefs than he did from the bottom. Mr. Hunter however adds, that he could ftill hear the cry, and diftinguifh the bird.

The

The Scotch Ptarmigan frequents the higheft ground of any Britifh bird, and he takes but very fhort flights.

But it is alfo urged by fome, that the reafon why feamen do not regularly fee the migration of birds, is becaufe they choofe the night, and not the day, for the paffage *.

Now though it may be allowed, that poffibly birds may crofs from the coaft of Holland to the Eaftern coaft of England (for example) during a long night, yet it muft be dark nearly as long as it is within the Arctic circle to afford time for a bird to pafs from the Line to many parts of Europe, which Monf. de Buffon calculates, may be done in about eight or nine days †.

If the paffage happened in half the nights of the year, which have the benefit of moonlight, the birds would be difcovered by the failors almoft as well as in the day time; to which I muft add that feveral fuppofed birds of paffage (the Fieldfare in particular) always call when on their flight, fo that the feamen muft be deaf as well as blind, if fuch flocks of birds efcape their notice.

Other objections however remain to this hypo-thefis of a paffage during the night.

---

* Mr. Catefby fuppofes that they may thus pafs in the night time, to avoid birds of prey. Phil. Tranf. Abr. Vol. II. p. 887. But are not owls then ftirring?

On the other hand, if they migrate in the day time, kites, hawks, and other birds of prey, muft be very bad fportfmen not to attend (like Arabs) thefe large and periodical caravans.

† In the preface to the firft volume of his lately publifhed Ornithology, p. 32.

Ninety-

Moſt birds not only ſleep during the night, but are as much incapacitated from diſtinguiſhing objects well as we are, in the abſence of the ſun: it is therefore inconceivable that they ſhould chooſe owl-light for ſuch a diſtant journey.

Beſides this, the Eaſtern coaſt of England, to which birds of paſſage muſt neceſſarily firſt come from the continent, hath many light-houſes upon it; they would therefore, in a dark night, immediately make for ſuch an object, and deſtroy themſelves by flying with violence againſt it, as is well known to every bat-fowler.

Having endeavoured to anſwer theſe two ſuppoſitions, by which it is contended that birds of paſſage may eſcape obſervation in their flight; I ſhall now conſider all the inſtances I have been able to meet with of any birds being actually ſeen whilſt they were croſſing any extent of ſea, though I might give a very ſhort refutation to them, by inſiſting, that if this was ever experienced, it muſt happen as conſtantly in a ſea, which is much navigated, as the return of the ſeaſons.

I cannot do better than to follow theſe according to chronological order.

The firſt in point of time is that which is cited by Willoughby *, from Bellon, whoſe words are thus tranſlated, " When we ſailed from Rhodes to " Alexandria, many quails flying from the North " towards the South, were taken in our ſhip, whence " I am perſuaded that they ſhift places; for for- " merly, when I ſailed out of the Iſle of Zant to " Morea, or Negropont, in the ſpring, I had ob-

* B. II. c. 11. §. 8.

" ſerved

" ferved quails flying the contrary way to N. and S.
" that they might abide there all fummer, at which
" time alfo a great many were taken in the fhip."

Let us now confider what is to be inferred from
this citation.

In the firft place, Bellon does not particularize the
longitude and latitude of that part of the Mediterranean, which he was then croffing; and in his courfe
from Rhodes to Alexandria, both the iflands of
Scarpanto and Crete could be at no great diftance :
thefe quails therefore were probably flitting from one
ifland of the Mediteranean * to another.

The fame obfervation may be made with regard
to the quails which he faw between Zant and Negropont, as the whole paffage is crouded with iflands,
they therefore might be paffing from ifland to ifland,
or headland to headland, which might very probably lye Eaft and Weft, fo as to occafion the birds
flying in a different direction, from which they paffed
the fhip before.

I have therefore no objection to this proof of migration, if it is only infifted upon to fhew that a quail
fhifts its ftation at certain feafons of the year; but
cannot admit that it is fair from hence to argue that
thefe birds periodically crofs large tracts of fea.

Bellon himfelf ftates, that when the birds fettled
upon the fhip, they were taken by the firft perfon
who chofe to catch them, and therefore they muft
have been unequal to the fhort flight which they
were attempting.

---

* One of the Mediterranean iflands is fuppofed to have obtained its ancient name of Ortygia from the numbers of quails.

It

It is very true that quails have been often pitched upon as instances of birds that migrate across seas, because they are scarcely ever seen in winter: it is well known, however, to every sportsman, that this bird never flies 300 yards at a time, and the tail being so short, it is highly improbable they should be equal to a passage of any length.

We find therefore, that quails, which are commonly supposed to leave our island in the winter, in reality retire to the sea coasts, and pick up their food amongst the sea weeds *.

I have happened lately to see a specimen of a particular species of quail, which is described by Dr. Shaw †, and is distinguished from the other kinds by wanting the hind-claw.

Dr. Shaw also states that it is a bird of passage. Now if quails really migrate from the coast of Barbary to Italy, as is commonly supposed, whence can it have arisen that this remarkable species hath escaped the notice of Aldrovandus, Olina, and the other Italian ornithologists?

When I had just finished what I have here said with regard to the migration of quails, I have had an opportunity of seeing the second volume of Monf. de Buffon's ornithology ‡ ; where, under this article, he contends that this bird leaves Europe in the winter.

It is incumbent upon me, therefore, either to own I am convinced by what this most ingenious and able naturalist hath urged, or to give my reasons why I

* See Br. Zool. Vol. II. p. 210. 2d Ed. octavo.
† Phyf. Obf. on the kingdom of Algiers, ch. 2.
‡ See p. 459, & feq.

ſtill continue to diſſent from the opinion he main-tains.

Though M. de Buffon hath diſcuſſed this point very much at large, yet I find only the following faƈts or arguments to be new.

He firſt cites the Memoirs of the Academy of Sciences *, for an account given by M. Godeheu of quails coming to the iſland of Malta in the month of May, and leaving it in September.

The firſt anſwer to this obſervation is, that the iſland of Malta is not only near to the coaſt of Africa, but to ſeveral of the Mediterranean iſlands ; it therefore amounts to no more than the flitting I have before taken notice of †.

Monſ. de Buffon ſuppoſes that a quail only quits one latitude for another, in order to meet with a perpetual crop on the ground.

Now can it be ſuppoſed that there is that difference between the harveſt on the coaſt of Africa, and that of the ſmall quantity of grain which grows on the rocky iſland of Malta, that it becomes inconvenient to the bird to ſtay in Africa as ſoon as May ſets in ; and neceſſary, on the other hand, to continue in Malta from May till September.

Monſ. de Buffon then ſuppoſes that quails make their paſſage in the night, as well as conceives them to be of a remarkably warm temperature ‡, and ſays

* Tom. III. p. 91 and 92.

† Both Monſ. de Godeheu and M. de Buffon ſeem to conceive that the quail ſhould fly in the ſame direƈtion as the wind blows ; but birds on the wing from point to point, which are at a conſiderable diſtance, fly againſt the wind, as their plumage is otherwiſe ruffled.

‡ As this is given for a reaſon why the African quails migrate Northward : Q. what is to become of the Icelandic quails during the ſummer?

that " *chaud comme une caille,*" is in every one's mouth *.

Now in the firſt place their migration during the night, is contrary to Belon's account, which M. de Buffon ſo much relies upon, who expreſly ſays, that the birds were caught in the day time †.

In the next place, I apprehend that " *chaud comme une caille,*" alludes to the very remarkable falacioufnefs of this bird, and not to the conſtant heat of its body.

Monſ. de Buffon then obferves, that if quails are kept in a cage, they are remarkably impatient of confinement in the autumn and ſpring, whence he infers that they then want to migrate ‡; he alſo adds, in the ſame period, that this uneaſineſs begins an hour before the ſun riſes, and that it continues all the night.

This great naturaliſt does not ſtate this obſervation as having been made by himſelf, and it ſeems upon the face of it to be a very extraordinary one.

---

* All birds indeed are warmer by four degrees than other animals. See ſome ingenious thermometrical experiments by Mr. Martin of Aberdeen, Edinb. 1771, 12mo.

† Upon looking a ſecond time into Belon, he does not indeed ſtate whether it was in the day or the night; but if it had happened in the latter, this traveller and ornithologiſt could not well have omitted ſuch a circumſtance. Beſides this, he mentions in what direction the quails were flying, which he could not have diſcerned in the night.

‡ It may alſo ariſe from this bird's being of ſo quarrelſome a difpoſition, and confequently moſt likely to fight with its fellow priſoners when they are all in greateſt vigour after moulting, and on the return of the ſpring.

M. de Buffon allows that they will fight for a grain of millet, and adds, " car parmi les animaux il faut un ſujet reel pour ſe " battre." M. de Buffon hath never been in a cockpit.

No

No one (at leaft with us) ever keeps quails in a cage except the poulterers, who always fell them as faft as they are fat, and confequently can give no account of what happens to them during fo long an imprifonment as this obfervation neceffarily implies.

No fuch remarkable uneafinefs hath ever been attended to in any other fuppofed bird of paffage during its confinement; but, allowing the fact to be as M. de Buffon ftates, he himfelf fupplies us with the real caufe of this impatience.

He afferts, that quails conftantly moult twice * a year, viz. at the clofe both of fummer and winter; whence it follows, that the bird, in autumn and the fpring, muft be in full vigour upon its recovery from this periodical illnefs: it can therefore as little brook confinement, as the phyfician's patient upon the return of health after illnefs.

Thus much I have thought it neceffary to fay, in anfwer to M. de Buffon, who "dum errat, docet," who fcarcely ever argues ill but when he is mifinformed as to facts, and who often, from ftrength of underftanding, difbelieves fuch intelligence as might impofe upon a naturalift of lefs acutenefs and penetration.

---

* I have often heard that certain birds moult twice a year, fome of which I have kept myfelf without their changing their feathers more than once.

I fhould fuppofe that this notion arifes from fome birds not moulting regularly in the autumn every year; and when the change takes place in the following fpring, they very commonly die: I can fcarcely think that many of them are equal to two illneffes of fo long a continuance, which are conftantly to return within twelve months.

I fhould therefore rather account for the extraordinary brifknefs of a quail in autumn and the fpring, from its recovery after moulting in the former, and from the known effects of the fpring as to moft animals in the latter.

The

The next inftance of a bird being caught at any diftance from land, is in Sir Hans Sloane's voyage to Jamaica, who fays, that a lark was taken in the fhip 40 leagues from the fhore: this therefore was certainly an unfortunate bird, forced out to fea by a ftrong wind in flying from headland to headland, as no one fuppofes the fkylark to be a bird of paffage.

The fame anfwer may be given to a yellow-hammer's fettling upon Haffelquift's fhip in the entrance of the Mediterranean, with this difference, that either the European or African coaft muft have been much nearer than 40 leagues *.

The next fact to be confidered is what is mentioned in a letter of Mr. Peter Collinfon's, printed in the Philofophical Tranfactions †.

He there fays, " That Sir Charles Wager had
" frequently informed him, that in one of his
" voyages home in the fpring as he came into found-
" ings in our channel, that a great flock of fwallows
" almoft covered his rigging, that they were nearly
" fpent and famifhed, and were only feathers and
" bones ; but being recruited by a night's reft, they
" took their flight in the morning."

The firft anfwer to this is, that if thefe were birds which had croffed large tracts of fea in their periodical migrations, the fame accident muft happen eternally, both in the fpring and autumn, which is not however pretended by any one.

In the next place, the fwallows are ftated to be fpent both by famine and fatigue; and how were they to procure any flies or other fuftenance on the

* See Haffelquift's Travels, in princ.
† 1760, Part II. p. 461.

3   rigging

rigging of the admiral's ſhip, though they migth indeed reſt themſelves?

Sir Charles, however, expreſly informs us, that he was in the channel, and within ſoundings: theſe birds, therefore (like Bellon's quails) were only paſſing probably from headland to headland; and being forced out by a ſtrong wind, were obliged to ſettle upon the firſt ſhip they ſaw, or otherwiſe muſt have dropped into the ſea, which I make no doubt happens to many unfortunate birds under the ſame circumſtances.

As the birds which thus ſettled upon Sir Charles Wager's rigging were ſwallows, it very naturally brings me now to conſider the celebrated obſervation of Monſ. Adanſon, under all its circumſtances, as it hath been ſo much relied upon, and by naturaliſts of ſo great eminence.

Monſ. Adanſon is a very ingenious writer, and the publick is much indebted to him for many of the remarks which he made whilſt he reſided in Senegal.

I may, however, I think, preſume to ſay, that he had not before his voyage made ornithology his particular ſtudy; proofs of which are not wanting in other parts of his work, which do not relate to ſwallows.

For example, he ſuppoſes, that the Canary birds which are bred in Europe are white, and that they become ſo by our climate's being more cold than that of Africa.

" J'ai remarqué que le ſerin qui devient tout blanc
" en France, eſt a Teneriffe d'un gris preſque auſſi
" foncé que celui de la linotte; ce changement de
" couleur provient vraiſemblablement de la froidure
" de notre climat *."

* Voyage au Senegal, p. 13.

Mr.

Mr. Adanſon in this paſſage ſeems to have deduced two falſe inferences from having ſeen a few white Canary birds in France, which he afterwards compares with thoſe of Teneriff, and ſuppoſes the change of colour to ariſe merely from alteration of climate : it is known, however, almoſt to every one, that there is an infinite variety in the plumage of the European Canary birds, which, as in poultry, ariſes from their being pampered with ſo much food, as well as confinement *.

Monſ. Adanſon, in another part of his voyage †, deſcribes a Roller, which he ſuppoſes to migrate ſometimes to the Southern parts of Europe.

This circumſtance ſhews that he could not have looked much into books of natural hiſtory, becauſe the principal ſynonym of this bird is *garrulus Argentoratenſis* ‡ ; and Linnæus informs us that it is found even in Sweden ‖.

---

* In the ſame paſſage, he compares the colour of the African Canary bird to that of the European linnet, and ſays it is *d'un gris preſque auſſi foncé*, whereas the European linnet is well known to be brown, and not grey. The linnet affords a very deciſive proof that the change of plumage does not ariſe from the difference of climate, but the two cauſes I have aſſigned. The cock bird, whilſt at liberty, hath a red breaſt : yet if it is either bred up in a cage from the neſt, or is caught with its red plumage, and afterwards moults in the houſe, it never recovers the red feathers.

That moſt able naturaliſt, Monſ. de Buffon, from having ſeen ſome cock linnets which had thus moulted off, or perhaps ſome hen linnets (which have not a red breaſt) conſiders them as a diſtinct ſpecies, and compares their breeding together in an aviary, to that of the Canary bird and goldfinch. Ornith. p. XXII.

† P. 16.   ‡ Or of Straſburgh.

‖ Faun. Suec. 94.

The

The ftrong characteriftic mark of this bird, is the outermoft feathers of the tail, which able naturalifts defcribe as three fourths of an inch longer than the reft *. Monf. Adanfon, however, compares their length, not with the other feathers of the tail, but with the length of the bird's body, which is by no means the natural or proper ftandard of comparifon.

The reafon of my taking notice of thefe more minute inaccuracies in Monf. Adanfon's account of birds, arifes from Mr. Collinfon's relying upon his obfervations with regard to fwallows being fo abfolutely decifive, becaufe he is reprefented to be fo able a naturalift.

I fhall now ftate (very minutely) under what circumftances thefe fwallows were caught, and what feems to be the true inference from his own account.

He informs us, that four fwallows fettled upon the fhip, not 50 leagues from the coaft of Senegal, on the 6th of October; that thefe birds were taken, and that he knew them to be the true fwallow of Europe†, which he fuppofes were then returning to the coaft of Africa.

I fhall now endeavour to fhew that thefe birds could not be European fwallows; nor, if they were, could they have been on their return from Europe to Africa.

* Willoughby, p. 131. Br. Zool. Vol. II. in append.
† I have before endeavoured to fhew that Monf. Adanfon does not always recollect with accuracy the plumage of the moft common European birds, by what he fays with regard to the linnet.

The

The word *hirondelle,* in French, is ufed as a general term for the four * fpecies of thefe birds, as the term *fwallow* is with us.

Now the four fwallows thus caught and examined by Monf. Adanfon were either all of the fame fpecies, or intermixed in fome other proportion.

Would not then any naturalift in ftating fo material a fact (as he himfelf fuppofes it to be) have particularized of what fpecies of fwallow thefe very interefting birds were?

Should not Monf. Adanfon alfo have taken care to diftinguifh thefe fuppofed European fwallows from two fpecies of the fame tribe, which bear a general refemblance to thofe of Europe, and are not only defcribed, but engraved by Briffon, under the name of *Hirondelle de Senegal & Hirondelle de rivage du Senegal* + ?

Though Monf. Adanfon was above a year on this part of the African coaft, paid fo much attention to fwallows, and was fo immediately acquainted with the different fpecies on the firft infpection, yet he feems never to have difcovered that there were fuch African fwallows as are thus defcribed and engraved by Briffon, though he muft have feen them daily.

Monf. Adanfon however concludes his account of the fuppofed European fwallow, whilft it continues on the coaft of Senegal, by a circumftance which

---

* *Viz.* the fwallow καῖ᾽ εξοχην, the martin, the fand martin, and the fwift: I omit the goatfucker, becaufe this bird, though properly claffed as a fpecies of fwallow by ornithologifts, is not fo confidered by others.

+ See Briffon, Tom. II. pl. xiv.

feems to prove to demonftration of what fpecies the four fwallows caught in the fhip really were.

He fays that they rooft on the fand either by themfelves, or at moft only in pairs, and that they frequent the coaft much more than the inland parts *.

Thefe fwallows therefore, if they came from Europe, muft have immediately changed at once their known habits : and is it not confequently moft clear that they were of that fpecies which Briffon defcribes under the name of *Hirondelle de rivage du Senegal?*

But though it fhould be admitted, notwithftanding what I have infifted upon, from Monf. Adanfon's own account, that thefe were really fwallows of the fame kind with thofe of Europe; yet I muft ftill contend that they could not poffibly have been on their return from Europe to Africa, becaufe the high road for a bird from the moft Weftern point of Europe to Senegal, is along the N. Weft coaft of Africa, which projeƈts greatly to the Weftward of any part of Europe.

What then could be the inducement to thefe four fwallows to fly 50 leagues to the Weftward of the coaft of Senegal, fo much out of the proper direƈtion?

It feems to me therefore, very clear, that thefe fwallows (whether of the European kind or not) were flitting from the cape de Verde iflands to the

* Voyage au Senegal, p. 67. I wifh Monf. Adanfon had alfo informed us whether thefe fwallows had the fame notes with thofe of Europe, which is a very material circumftance in the natural hiftory of birds, though little attended to by moft orni-thologifts.

coaft of Africa, to which fhort flight, however, they were unequal, and were obliged from fatigue to fall into the failors hands.

Monf. Adanfon likewife mentions * that the fhip's company caught a Roller on the 26th of April, which he fuppofes was on its paffage to Europe, though he was then within fight of the coaft of Senegal : this bird, however, muft be admitted not to have had fufficient ftrength to reach the firft ftage of this round-about journey, and was therefore probably forced out to fea by a ftrong wind, in paffing from head-land to head-land.

But I muft not difmifs what hath been obferved with regard to the fwallows feen by Monf. Adanfon at Senegal, without endeavouring alfo to anfwer what M. de Buffon hath not only inferred from it, but hath endeavoured to confirm by an actual experiment †.

M. de Buffon, from the many inftances of fwallows being found torpid even under water, very readily admits, that all the birds of this genus do not migrate, but only that fpecies which was feen by Monf. Adanfon in Africa, and which he generally refers to as the chimney fwallow ‡; but from the outfet, feems

---

* Voyage au Senegal, p. 15.

† See the two prefatory difcourfes to his fixteenth volume of natural hiftory.

‡ So little do naturalifts know of this very common bird, that I believe it hath never yet been obferved by any writer, that the male fwallow hath only the long flender feathers in the tail, which are confidered as its moft diftinguifhing marks. I venture to make this remark upon having feen the difference in two fwallows which are in Mr. Tunftall's collection, F. R. S. as alfo in two others, which have lately been prefented to the Mufeum

to

to shew that he hath himself confounded this species with the martin.

" Prenons un seul oiseau, par exemple, l'hiron-
" delle, celle que tout le monde connoit, qui paroit
" au printems, disparoit en automne, & fait son nid
" avec de la terre contre les fenetres, ou dans les
" cheminees." p. 23.

It is very clear that the design in this period is to specify a particular bird in such a manner that no doubt could remain with any one about the species referred to; and from other passages which follow, it is as clear that Monf. de Buffon means to allude to the swallow κατ' εξοχην.

Though this was certainly the intention of this most ingenious naturalist, it is to me very evident that the martin, and not the swallow, was in his contemplation, because he first speaks of the bird's building against windows, before he mentions chimneys, and therefore supposes that either place is indifferent ; which is not the case, because the swallow seldom builds on the sides of windows, or the martin in chimneys.

There are perhaps three or four martins to one swallow in all parts; and from their being the more common bird of the two, as well as from the circumstance of their building at the corner of windows (and consequently being eternally in our fight), nine-

of the Royal Society, by the directors of the Hudson's Bay company.
These long feathers would be very inconvenient to the hen during incubation ; and they are likewife confined to the cock *widow-bird*, as, from their more extraordinary length, they would be still more so.

teen

teen out of twenty, when they fpeak of a fwallow, really mean a martin *.

I only take notice of this fuppofed inacuracy in Monf. de Buffon, becaufe, if that able naturalift does not fpeak of the different forts of fwallows with that precifion which is neceffary upon fuch an occafion, why fhould he rely fo intirely upon the impoffibility of Monf. Adanfon's being miftaken?

I fhall now ftate the experiment of Monf. de Buffon, to prove that the fwallow is not torpid in the winter, and muft therefore migrate to the coaft of Senegal †.

He fhut up fome fwallows *(hirondelles)* in an ice houfe, which were there confined " plus ou moins " de temps;" and the confequence was, that thofe which remained there the longeft died, nor could they be revived by expofing them to the fun; and, that thofe " qui n'avoient fouffert le froid de la " glaciere que pendant peu de tems" were very lively when permitted to make their efcape.

* In the fame manner the generical name in other languages, for this tribe of birds, always means the martin, and not the fwallow.

Thus Anacreon complains of the χελιδων for waking him by its twittering.

Now if it be confidered that there was only the kitchen chimney in a Grecian houfe, it muft have been the martin which built under the eaves of the window, that was troublefome to Anacreon, and not the fwallow.

Ovid alfo fpeaking of the neft of the *hirundo*, fays,

—— luteum fub trabe figit opus.

by which he neceffarily alludes to the martin, and not the fwallow.

† Plan de l'ouvrage, p. 15.

Monf.

Monf. de Buffon does not, in this account of his experiment, ftate the time during which the birds were confined; but as the trial muft have been made in France, the fwallows which he procured could not be expected to be torpid either in an ice-houfe * or any other place, becaufe the feafon for their being in that ftate was not yet arrived.

I cannot alfo agree with M. de Buffon that thofe birds which were fhut up the longeft time died through cold, as he fuppofes, but for want of food, as he neither fupplied them with any flies, nor, if he had, could the fwallows have caught them in the dark : a very fhort faft kills thefe tender animals, which are feeding every inftant when on the wing.

It therefore feems not to follow from this, or any other experiment, that fwallows muft neceffarily migrate (as Monf. de Buffon fuppofes) to the coaft of Senegal.

* The very name of an ice-houfe almoft ftrikes one with a chill; I placed, however, a thermometer in one near Hyde Park Corner, on the 23d of November, where it continued 48 hours, and the mercury then ftood at $43\frac{1}{2}$ by Fahrenheit's fcale.

This is therefore a degree of cold which fwallows fometimes experience whilft they continue in fome parts of Europe, without any apparent inconvenience ; and it fhould feem that the cold vapours which may arife from the included ice, fink the ther- mometer only 7 or 8 degrees, as the temperature in approved cellars is commonly from 50 or 51 throughout the year.

Sir William Hamilton informs me, that he hath frequently feen fwallows in the winter between Naples and Puzzuoli, when the weather was warm ; as does Mr. Hunter, F. R. S. that he hath obferved them during the fame feafon, on the confines of Spain and Portugal. It fhould feem from this, that very mild and warm weather for any continuance always wakes thefe birds from their ftate of torpidity.

Swallows

Swallows are feen during the fummer, in every part of Europe from Lapland to the Southern coaft of Spain; nor is Europe vaftly inferior in point of fize to Africa.

If fwallows therefore retreat to Africa in the winter, fhould not they be difperfed over the whole Continent of Africa, juft as they are over every part of Europe?

But this moft certainly is not fo: Dr. Shaw, who was a very good naturalift and attended much to the birds in the neighbourhood of Algiers (as appears by his account of that country), makes no mention of any fuch circumftance, nor have we heard of it from any other traveller *.

It muft be admitted indeed, that Herodotus fpeaking of a part of upper Egypt (which he had never feen) fays, that kites and fwallows never leave it †; this, however, totally differs from Monf. Adanfon's account, who informs us that they difappear in Senegal on the approach of fummer.

It feems to follow therefore, from this filence in others, that fwallows cannot be accommodated for their winter refidence in any part of that vaft continent, but in the neighbourhood of Senegal.

But this is not the whole objection to fuch an hypothefis.

---

* It may alfo be obferved here, that credit is in fome meafure given to M. Adanfon's eyefight, againft that of all the Englifh, French, Dutch, Portugueze, and Danes, who have been fettled not far from Senegal for above a century, many of which have fpent the greateft part of their lives there, and whofe notice, fwallows feen during the winter, muft have probably attracted.

† Ικτινοι δε και χελιδονες δι ετεος εοντες ουκ απολειπουσι. Euterpe, p. 98. ed. Gale.

If

If the fwallows of Europe, when they difappear in thofe parts, retreat to the coaft of Senegal, what neceffarily follows with regard to a Lapland fwallow ?

I will fuppofe fuch a bird to have arrived fafely at his winter quarters upon the approach of that feafon in Lapland; but he muft then, according both to Monf. Adanfon's and de Buffon's account, return to Lapland in the fpring, or at leaft fome other fwallow from Senegal fill his place *.

Such a bird immediately upon its arrival on the Southern coaft of Spain would find the climate and food which it defired to attain, and all proper conveniences for its neft : what then is to be its inducement for quitting all thefe accommodations which it meets with in fuch profufion, and pufhing on immediately over fo many degrees of European continent to Lapland, where both martin and fwallow can procure fo few eaves of houfes to build upon ? What alfo is to be the inducement to thefe birds, when they have arrived at that part of the Norwegian coaft which is oppofite to the Ferroe iflands, to crofs degrees of fea, in order

---

* Mr. Stephens, A. S. S. informs me, that there was a neft of martins for twenty years together in the hall of his houfe in Somerfetfhire (near Bath) ; nor could the old birds procure food either for themfelves, or their young, till the door was opened in the morning.

Can it it be fuppofed that the fame birds or their defcendants could have fo long fixed upon fo very inconvenient a fpot, to which they conftantly returned from the coaft of Africa, neglecting fo many others, which they muft have always paffed by ? Does it not alfo afford a moft ftrong prefumption, that they were torpid during winter in the neighbourhood of this old hall ?

to

to build in fuch fmall fpots of land, where there are ftill fewer houfes?

The next fact I have happened to meet with of a bird's being feen at a confiderable diftance from the fhore, is in Mr. Forfter's lately publifhed tranflation of Kalm's account of N. America*.

We are there informed that a bird (which Kalm calls a fwallow) was feen near the fhip on the 2d of September, and, as he fuppofes, 20 degrees from the continent of America †.

It appears, however, by what he before ftates in his journal, that the fhip was not above 5 degrees from the ifland of Sable.

Befides, if it is contended that this was an European fwallow on its paffage acrofs the Atlantic on the 2d of September, it is too early even for a fwift, to have been on its migration, which difappears with us fooner than the three other fpecies of European fwallows ‡.

Only two more inftances have occurred of birds being feen in *open* fea that have been defcribed

* Vol. I. p. 24.

† It may not be improper here to obferve, that in all inftances of birds being feen at fea any great diftance from the coaft, it is not improbable that they may have before fettled on fome other veffel, or perhaps on a piece of floating wreck.

By accidents of this fort, even butterflies have fometimes been caught by the failors at 40 leagues diftance from any land. See Monf. l'Abbé Courte de la Blanchadiere's Voyage to Brazil, Paris, 1759, 21mo. p. 169.

‡ The bird mentioned by Kalm was probably an American fwallow, forced out to fea by fome accidental ftorm : there are feveral fpecies of them and they feem to bear a general affinity to thofe of Europe.

with

with any fort of precifion, which I fhall juft
ftate, as I would not decline giving the beft anfwer
I am able to every argument and fact which may be
relied upon, by thofe who contend that birds periodically migrate acrofs oceans.

On the 30th of March, 1751, Ofbeck, in his
voyage from Sweden to China *, met with a fingle
houfe fwallow near the Canary Iflands, which was
fo tired that it was caught by the failors: Ofbeck
alfo ftates, that though it had been fine weather for
feveral preceding days, the bird was as wet as if it
had juft emerged from the bottom of the fea.

If this inftance proves any thing, it is the fub-
merfion and not the migration of fwallows fo generally believed in all the northern parts of Europe.
It would fwell this Letter to a moft unreafonable
fize, to touch only upon this litigated point ; and I
fhall, for the prefent, fupprefs what hath happened
to occur to me on this controverted queftion †.

---

* See the lately publifhed tranflation of this voyage.

† I will, however, mention one moft decifive fact on this
head.

Mr. Stephens, A. S. S. informs me, that, when he was
fourteen years of age, a pond of his father's (who was vicar of
Shrivenham in Berkfhire) was cleaned, during the month of
February ; that he picked up himfelf a clufter of three or four
fwallows (or martins), which were caked together in the mud,
and that he carried them into the kitchen, on which they foon
afterwards flew about the room, in the prefence of his father,
mother, and others. Mr. Stephens alfo told me, that his father
(who was a naturalift) obferved at the time, he had read of fimilar
inftances in the northern writers. This fact is alfo confirmed to
me by the Reverend Dr. Pye, who was then at fchool in Shrivenham, as alfo by a very fenfible land-furveyor, who now lives
in the village.

Ofbeck afterwards, in the courfe of his voyage, mentions, that a fwallow (indefinitely) followed the fhip, near Java, on the 24th of July, and another on the 14th of Auguft, in the Chinefe fea, as he terms it.

After what I have obferved before with regard to other inftances of the fame fort, I need fcarcely fay that this naturalift does not ftate of what fpecies thefe fwallows were; and that, from the latitudes in which they were feen, they muft have been fome of the Afiatic kinds.

I cannot, however, difmifs this article of the fwallow, without adding fome general reafons, which feem to prove the great improbability of this or any other bird's periodically migrating over wide tracts of fea; and I the rather do it in this place, becaufe

There are feveral reafons why fwallows fhould not be frequently thus found; ponds are feldom cleaned in the winter, as it is fuch cold work for the labourers; and the fame inftinct which prompts the bird thus to conceal itfelf, inftructs it to choofe fuch a place of fecurity, that common accidents will not difcover it.

But the ftrongeft reafon for fuch accounts not being more numerous, is, that facts of this fort are fo little attended to; for though I was born within half a mile of this pond, and have always had much curiofity with regard to fuch facts, yet I never heard a fyllable about this very material and interefting account, till very lately.

To this fact I muft alfo add, that fwallows may be conftantly taken in the month of October, during the dark nights, whilft they fit on the willows in the Thames, and that one may almoft inftantaneoufly fill a large fack with them, becaufe at this time they will not ftir from the twigs, when you lay your hands upon them. This looks very much like their beginning to be torpid before they hide themfelves under the water.

A man near Brentford fays, that he hath caught them in this ftate in the eyt oppofite to that town, even fo late as November.

the

the fwallow is commonly pitched upon as the moft notorious inftance of fuch a regular paffage.

This feems to arife firft from its being feen in fuch numbers during the fummer, from its appearing almoft always on the wing, and from its feeding in that pofition; from which two latter circumftances it is fuppofed to be the beft adapted for fuch diftant migrations.

And firft, let us confider, from the few facts or reafons we have to argue from, what length of flight either a fwallow or any other bird is probably equal to.

A fwallow, it is true, feems to be always on the wing; but I have frequently attended, as much as I could, on a particular one; and it hath appeared to me, that the bird commonly returned to its neft in eight or ten minutes: as for extent of flight, I believe I may venture to fay, that thefe birds are feldom a quarter of mile from their mate or young ones; they feed whilft on the wing, and are perpetually turning fhort round to catch the infects, who endeavour to elude them as a hare does a greyhound.

It therefore feems to me, that fwallows are by no means equal to long flights, from their practice during their fummer refidence with us.

I have long attended to the flight of birds; and it hath always appeared to me, that they are never on the wing for amufement (as we walk or ride), but merely in fearch of food.

The only bird which I have ever obferved to fly without any particular point of direction, is the rook: thefe birds will, when the wind is high,

" Ride

" Ride in the whirlwind, and enjoy the ftorm."

They never fly, however, at this time, from point to point, but only tumble in the air, merely for their diverfion.

It feems, therefore, that birds are by no means calculated for flights acrofs oceans, for which they have no previous practice: and they are, in fact, always fo fatigued, that, when they meet a fhip at fea, they forget all apprehenfions, and deliver themfelves up to the failors.

Let us now confider another objection to the migration of the fwallow, which Monf. de Buffon fuppofes may crofs the Atlantic to the Line in eight days * ; and this not only from the want of reft, but of food, during the paffage.

A fwallow, indeed, feeds on the wing: but where is it to find any infects, whilft it is flying over a wide expanfe of fea ? This bird, therefore, if it ever attempted fo adventurous a paffage, would foon feel a want of food, and return again to land, where it had met with a conftant fupply from minute to minute.

I am aware it may be here objected, that the fwallow leaves us on the approach of winter, when foon no flying infects can be procured: but I fhall hereafter endeavour to fhew, that thefe birds are then torpid, and, confequently, can want no fuch food.

Another objection remains to the hypothefis of migration, which is, that birds, when flying from

* Difcours fur la nature des oifeaux, p. 32.

point

point to point, endeavour always to have the wind
againſt them *, as is periodically experienced by the
London bird-catchers, in March and October, when
they lay their nets for ſinging birds †.

The reaſon, probably, for birds thus flying againſt
the wind is, that their plumage may not be ruffled,
which indeed I have before had occaſion to mention.

Let us ſuppoſe, then, a ſwallow to be equal to a
paſſage acroſs the Atlantic in other reſpects; how is
the bird to be inſured of the wind's continuing for
days in the ſame quarter; or how is he to depend
upon its continuing to blow againſt his flight with
moderation? for who can ſuppoſe that a ſwallow can
make his way to the point of direction, when buf-
feted by a ſtorm blowing in the teeth of his intended
paſſage ‡?

Laſtly, can it be conceived that theſe, or any
other birds, can be impelled by a providential inſtinct, regularly to attempt what ſeems to be attended with ſuch inſuperable difficulties, and what
moſt frequently leads to certain deſtruction?

But it will ſtill be objected, that as ſwallows regularly appear and diſappear at certain ſeaſons, it is
incumbent upon thoſe who deny their migration, to

* Kalm, in his voyage to America, makes the ſame obſervation, with regard to flying fiſh, and Valentine ſays, that if
the wind does not continue to blow againſt the bird of paradiſe,
it immediately drops to the ground.

† Theſe birds, as it ſhould ſeem, are then in motion; becauſe, at thoſe ſeaſons, the ground is plowed either for the winter
or lent corn.

‡ I have myſelf attended to ſwallows during a high wind,
and have obſerved that they fly only in ſheltered places, whilſt
they almoſt touch the ſurface of the ground.

ſhew

fhew what becomes of them in Europe during our winter.

Though it might be anfwered, that it is not neceffary, thofe who endeavour to fhew the impoffibility of another fyftem or hypothefis, fhould from thence be obliged to fet up one of their own; yet I fhall, without any difficulty, fay, that I at leaft am convinced fwallows (and perhaps fome other birds) are torpid during the winter.

I have not, I muft own, myfelf ever feen them in this ftate; but, having heard inftances of their being thus found, from others of undoubted veracity, I have not fcarcely the leaft doubt with regard to this point.

It is, indeed, rather difficult to conceive why fome ornithologifts continue to withhold their affents to fuch a cloud of witneffes, except that it perhaps contradicts a favourite hypothefis which they have already maintained.

Why is it more extraordinary that fwallows fhould be torpid during the winter, than that bats are found in this ftate, and fo many infects, which are the food of fwallows?

But it may be faid, that as the fwallows have crowded the air during the fummer, in every part of Europe fince the creation, and as regularly difappear in winter, why have not the inftances of their being found in a torpid ftate been more frequent?

To this it may be anfwered, that though our globe may have been formed fo many centuries, yet the inhabitants of it have fcarcely paid any attention to the ftudy of natural hiftory, but within thefe late years.

As

As for the ancient Greeks and Romans, their drefs prevented their being fo much in the fields as we are; or, if they heard of a rather extraordinary bird in their neighbourhood, they had not a gun to fhoot it: the only method of attaining real knowledge in natural hiftory, depends almoft entirely upon the having frequent opportunities of thus killing animals, and examining them when dead.

If they did not ftir much in their own country, much lefs did they think of travelling into diftant regions; want of bills of exchange, and of that curiofity which arifes from our being thoroughly acquainted with what is near us at home, probably occafioned this; to which may alfo be added, the want of a variety of languages: fcarcely any Greek feems to have known more than his own tongue, nor Roman more than two *.

Ariftotle, indeed, began fomething like a fyftem of natural hiftory, and Pliny put down, in his common place-book, many an idle ftory; but, before the invention of printing, copies of their works could not be fo generally difperfed, as to occafion much attention to what might be interefting facts for the natural hiftorian.

In the fixteenth century, Gefner, Belon, and Aldrovandus, publifhed fome materials, which might be of ufe to future naturalifts; but, in the feventeenth, Ray and Willoughy firft treated this extenfive branch of ftudy, with that clearnefs of method,

---

* It need be fcarcely here mentioned alfo, that their navigation was confined to the Mediteranean, from the compafs not having been then difcovered.

perfpicuity of defcription, and accuracy of obfervation, as hath not, perhaps, been fince exceeded.

The works of thefe great naturalifts were foon difperfed over Europe, and the merit of them acknowledged; but it fo happened, that Sir Ifaac Newton's amazing difcoveries in natural philofophy making their appearance about the fame time, engaged entirely the attention of the learned.

In procefs of time, all controverfy was filenced by the demonftration of the Newtonian fyftem; and then the philofophical part of Europe naturally turned their thoughts to other branches of fcience.

Since this period, therefore, and not before, natural hiftory hath been ftudied in moft countries of Europe; and confequently, the finding fwallows in a ftate of torpidity, or on the coaft of Senegal, during the winter, begins to be an interefting fact, which is communicated to the world by the perfon who obferves it.

To this I may add, that the common labourers, who have the beft chance of finding torpid birds, have fcarcely any of them a doubt with regard to this point; and confequently, when they happen to fee them in this ftate, make no mention of it to others; becaufe they confider the difcovery as neither uncommon or interefting to any one.

Molyneux, therefore, in the Philofophical Tanf-actions *, informs us, that this is the general belief of the common people of Ireland, with regard to land-rails; and I have myfelf received the fame anfwer from a perfon who, in December, found fwallows torpid in the ftump of an old tree.

* Phil. Tranf. abr. Vol. II. p. 853.

Another

Another reafon why the inſtances of torpid ſwallows may not be expected ſo frequently, is, that the inſtinct of ſecreting themſelves at the proper ſeaſon of the year, likewiſe ſuggeſts to them, it's being neceſſary to hide themſelves in ſuch holes and caverns, as may not only elude the ſearch of man, but of every other animal which might prey upon them; it is not therefore by any common accident that they are ever diſcovered in a ſtate of torpidity.

Since the ſtudy of natural hiſtory, however, hath become more general, proofs of this fact are frequently communicated, as may appear in the Britiſh Zoology *.

That it may not be ſaid, however, I do not refer to any inſtance which deſerves credit, if properly ſifted, I beg leave to cite the letter from Mr. Achard to Mr. Collinſon, printed in the Philoſophical Tranſactions †, from whence it ſeems to be a moſt irrefragable fact, that ſwallows ‡ are annually diſcovered in a torpid ſtate on the banks of the Rhine. I ſhall alſo refer to Dr. Birch's Hiſtory of the Royal Society ||, where it is ſtated, that the celebrated Harvey diſſected

---

* See Vol. II. p. 250. Brit. Zool. ill. p. 13, 14. As alſo Mr. Pennant's Tour in Scotland, p. 199.

† 1763, p. 101.

‡ "Swallows or martins," are Mr. Achard's words, which I the rather mention, becauſe Mr. Collinſon complains that the ſpecies is not ſpecified.

Mr. Collinſon himſelf had endeavoured to prove, that ſand martins are not torpid, Phil. Tranſ. 1760, p. 109. and concludes his letter, by ſuppoſing that all the ſwallow tribe migrates, therefore the ſwift is the only ſpecies remaining; for his friend Mr. Achard ſhews to demonſtration, that ſwallows or martins are torpid; he does not, indeed, preciſely ſtate which of them.

|| Vol. IV. p. 537.

some, which were found in the winter, under water, and in which he could not observe any circulation of the blood *.

Assuming it, therefore, from these facts, that swallows have been found in such a state, I would ask the partisans of migration, whether any instance can be produced where the same animal is calculated for a state of torpidity and, at the same time of the year, for a flight across oceans ?

But it may be urged, possibly, that if swallows are torpid when they disappear, the same thing should happen with regard to other birds, which are not seen in particular parts of the year.

To this I answer, that this is by no means a necessary inference: if, for example, it should be insisted that other birds besides the cuckow are equally careless with regard to their eggs, it would be immediately allowed that the argument arising from

---

* As the swallows were found in the winter, they must have been in a state of torpidity, as otherwise the animals must have been putrid.

I shall likewise here refer to Phil. Transf. abr. Vol. V. p. 33. where Mr. Derham says, that he heard a swift squeak in an hole of his house on the 17th of April ; but that, the weather being cold, it did not stir abroad for several days.

This seems to be a strong instance of a bird's first waking from a state of torpidity, but resuming its sleep on the weather being severe.

I shall close the proofs on this head (which I could much enlarge) by the dignified testimony of Sigismond, King of Poland, who affirmed on his oath, to the cardinal Commendon, that he had frequently seen swallows, which were found at the bottom of lakes. See the life of cardinal Commendon, p. 211. Paris, 1671. 4to.

such supposed analogy could by no means be relied upon *.

It is possible, however, that some other birds, which are conceived to migrate, may be really torpid as well as swallows; and if it be asked why they are not sometimes also seen in such a state during the winter, the answer seems to be, that perhaps there may be a thousand swallows to any other sort of bird, and that they commonly are found torpid in clusters.

* I here suppose the common notion about the cuckow to be true ; because both learned and ignorant seem equally to agree in the fact.

During the present summer, however, a girl brought a full feathered young cuckow to a gentleman's house, where I happened to be, who said, that it had been for several days before fed by another bird of equal size with itself; which therefore could not be a hedge-sparrow, or other small bird, but the parent cuckow.

I have also lately been favoured, by Mr. Pennant, with the following extract from a manuscript of Derham's on instinct.

" The Rev. Mr. Stafford was walking in Glossop-dale in the
" Peak of Derbyshire, and saw a cuckow rise from its nest,
" which was on the stump of a tree, that had been some time
" felled, so as much to resemble the colour of the bird. In
" this nest were two young cuckows, one of which he
" fastened to the ground, by means of a peg and line, and very
" frequently, for many days, beheld the old cuckow feed these
" her young ones."

It is not impossible, therefore, that this most general opinion will turn out like the supposed effects of the venom of the tarantula ; and, indeed, it is difficult to conceive how so small a bird as a hedge-sparrow can feed a cuckow : it is also remarkable, that the witnesses often vary about the species of small bird thus employed.

It is possible, however, that the cuckow (though it may not hatch its young) may feed them, when grown too large for the foster parent.

If

If a fingle bird of any other kind happens to be feen in the winter, without motion or apparent warmth, it is immediately conceived that it died by fome common accident.

I fhall, however, without any referve, fay, that I rather conceive the notion which prevails with regard to the migration of many birds, may moft commonly arife from the want of obfervation, and ready knowledge of them, when they are feen on the wing, even by profeffed ornithologifts.

It is an old faying, that " a bird in the hand is " worth two in the bufh;" and this holds equally with regard to their being diftinguifhed, when thofe even who ftudy natural hiftory, have but a tranfient fight of the animal *.

If, therefore, a bird, which is fuppofed to migrate in the winter, paffes almoft under the nofe of a Linnæan, he pays but little attention to it, becaufe he cannot examine the beak, by which he is to clafs the bird. Thus I conceive, that the fuppofing a nightingale to be a bird of paffage arifes from not readily diftinguifhing it, when feen in a hedge, or on the wing †.

This bird is known to the ear of every one, by its moft ftriking and capital notes, but to the eye of very

* An ingenious friend of mine makes always a very proper diftinction between what he calls in-door and out-door naturalifts.

Thomas Willifel, who affifted Ray and Willughby much with regard to the natural hiftory of the animals of this ifland, never ftirred anywhere without his gun and fifhing-tackle.

† No two birds fly in the fame manner, if their motions are accurately attended to.

2   few

few indeed; becaufe the plumage is dull, nor is there any thing peculiar in its make.

The nightingale fings perhaps for two months *, and then is never heard again till the return of the fpring, when it is fuppofed to migrate to us from the continent, with redftarts, and feveral other birds.

That it cannot really do fo, feems highly probable, from the following reafons.

This bird is fcarcely ever feen to fly above twenty yards, but creeps at the bottom of the hedges, in fearch of maggots, and other infects, which are found in the ground.

If the fwallow is not fupplied with any food during its paffage acrofs oceans, much lefs can the nightingale be fo accommodated; and I have great reafon to believe, from the death of birds in a cage, which have had nothing to eat for twenty-four hours, that thefe delicate and tender animals cannot fupport a longer faft, though ufing no exercife at all.

To this I may alfo add, that thofe birds which feed on infects are vaftly more feeble than thofe whofe bills can crack feed, and confequently, lefs capable of bearing any extraordinary hardfhips or fatigue.

But other proofs are not wanting, that this bird cannot migrate from England.

* Whilft it fings even, the bird can feldom be diftinguifhed, becaufe it is then almoft perpetually in hedges, when the foliage is thickeft, upon the firft burft of the fpring, and when no infects can as yet have deftroyed confiderable parts of the leaves.

Nightin-

Nightingales are very common in Denmark, Sweden, and Ruſſia *, as alſo in every other part of Europe, as well as Aſia, if the Arabic name is properly tranſlated.

Now, if it is ſuppoſed that many of theſe birds which are obſerved in the ſouthern parts of England, croſs the German ſea, from the oppoſite coaſt of the continent; why does not the ſame inſtinct drive thoſe of Denmark to Scotland, where no ſuch bird was ever ſeen or heard + ?

But theſe are not all the difficulties which attend the hypotheſis of migration; nightingales are agreed to be ſcarcely ever obſerved to the weſtward of Dorſetſhire, or in the principality of Wales ‡, much leſs in Ireland.

I have alſo been informed, that theſe birds are not uncommon in Worceſterſhire, whereas they are exceſſively rare (if found at all) in the neighbouring county of Hereford.

Whence, therefore, can it ariſe, that this bird ſhould at one time be equal to the croſſing of ſeas, and at other times not travel a mile or two into an adjacent county? Does it not afford, on the other hand, a ſtrong proof, that the bird really continues

* See Dr. Birch's Hiſtory of the Royal Society, Vol. III. p. 189. Linnæi Fauna Suecica. and Biographia Britannica, art. FLETCHER; where it is ſaid, that they have in Ruſſia a greater variety of notes than elſewhere.

+ Sir Robert Sibbald, indeed, conceives the nightingale to be a bird of North Britain; but, if I can depend upon many concurrent teſtimonies, no ſuch bird is ever ſeen or heard ſo far northward at preſent, nor could I ever trace them in that direction further than Durham.

‡ I have, however, frequently ſeen the nightingale's congener (and ſuppoſed fellow-traveller) the redſtart in Wales.

on

on the fame fpot during the whole year, but happens not to be attended to, from the reafons I have before fuggefted?

I am therefore convinced, that if I was ever to live in the country during the winter, I fhould fee night-ingales, becaufe I fhould be looking after them, and I am accordingly informed, by a perfon who is well acquainted with this bird, that he hath frequently obferved them during this feafon *.

If it be afked, why the nightingales are all this time mute? the anfwer is, that the fame filence is experienced in many other birds, and this very mute-nefs is, in part the caufe why the bird is not attended to in winter.

I muft now afk thofe who contend for the migra-tion of a nightingale, what is to be its inducement for croffing from the continent to us? a fwallow, in-deed, may want flies in winter, if it ftays in Eng-land; but a nightingale is juft as well fupplied with infects on the continent, as it can be with us after its paffage †. I muft alfo afk, in what other part of

* I find they have alfo been feen in France during the winter. See a treatife, intitled, Aëdologue, Paris 1751. p. 23.

† I have omitted the mention of a more minute proof, that this bird cannot migrate from the continent, from the having kept them for fome years in a cage, and having been very attentive to their fong. Kircher (in his Mufurgia) hath given us the nightingale's notes in mufical characters, from which it appears that the fong of a German nightingale differs very materially from that of an Englifh one: now, if there was a communication by migration between the continent and England, the fong of thefe birds would not fo materially differ, as I may, perhaps, fhew, by fome ex-periments I have made, in relation to the notes of birds.

I have before mentioned, that Mr. Fletcher, who was embaf-fador from England to Ruffia in the time of Queen Elizabeth,

the

the world this bird is seen during the winter? must it migrate to Senegal with the swallow?

I am perfuaded likewife, that the cuckow never migrates from this ifland any more than the nightingale: this bird is either probably torpid in the winter, or otherwife is miftaken for one of the fmaller kind of hawks *; which it would be likewife in the fpring, was it not for its very particular note at that time, and which only lafts during courtfhip, as it does with the quail.

If there is fine weather in February, this bird fometimes makes this fort of call to its mate, whilft it is fuppofed to continue ftill on the continent.

An inftance is mentioned by Mr. Bradley †, of not only a fingle cuckow, but feveral, which were heard in Lincolnfhire, during the month of February; and that able naturalift Mr. Pennant informs me, another was heard near Hatcham in Shropfhire, on the 4th of February in the prefent year ‡.

---

observed that the fong of the Ruffian nightingale differed from that of the Englifh.

* Mr. Hunter, F. R. S. informs me, that he hath feen cuckows in the ifland of Belleifle during the winter, which is not fituated fo much to the fouthward, as to make it improbable that they may equally continue with us.

† Works of Nature, p. 77.

‡ Mr. Pennant received this account from Mr. Plimly, of Longnor in Shropfhire.

Thus likewife Mr. Edwards informs us, that the fea fowls near the Needles, which are commonly fuppofed to migrate in winter, appear upon the weather's being very mild. Effays, p. 197.

It

It is amazing how much the being interested to discover particular objects contributes to our readily distinguishing them.

I remember the being much surprized that a grey-headed game-keeper always saw the partridge on the ground before they rose, when I could not do the same. He told me, however, that the reason was, I lived in a time when the shooter had no occasion to give himself that trouble.

He then further explained himself, by saying, that when he was young, no one ever thought of aiming at a bird when on the wing, and consequently they were obliged to see the game before it was sprung. He added, that from this necessity he could not only distinguish partridges, but snipes and woodcocks, on the ground.

Another instance of the same kind, is the great readiness with which a person, who is fond of coursing, finds a hare sitting in her form : those, however, who are not interested about such sport, can scarcely see the hare, when it is under their nose, and pointed out to them.

But more apparent objects escape our notice, when we are not interested about them.

Ask any one, who hath not a botanical turn, what he hath seen in passing through a rich meadow, at the time it is most enamelled with plants in flower; and he will tell you, that he hath observed nothing but grass and daisies. If most gardeners even are in like manner asked whether the flowers of a bean grow on every side of the stalk, they will suppose that they do, whereas they, in reality, are only to be found on one fide.

The mouths of flounders are often turned different ways, which one would think could not well efcape the obfervation of the London fifhmongers; yet, upon afking feveral of them whether they had attended to this particular, I found they had not, till I fhewed them the proof in their own fhops.

A fifhmonger, however, knows immediately whether a fifh is in good eating order or not, on the firft infpection; becaufe this is a circumftance which interefts him.

I fhall, however, by no means fupprefs two arguments in favour of migration, which feem to require the fulleft anfwer that can be given to them.

The firft is, that there are certain birds, which appear during the winter, but difappear during the fummer; and it may be afked, where fuch birds can be fuppofed to breed, if they do not migrate from this ifland.

Thefe birds are in number four, viz. the fnipe, woodcock, redwing, and fieldfare.

As for the fnipe, I have a very fhort anfwer to give to the objection, as far as it relates to this bird; becaufe it conftantly breeds in the fens of Lincoln-fhire, Wolmar foreft, and Bodmyn downs; it is therefore highly probable, that it does the fame in almoft every county of England.

I muft own, however, that, till within thefe few years, I conceived the neft of a fnipe was as rarely feen in England, as that of a woodcock or fieldfare; and that able ornithologift Mr. Edwards fuppofes this to

be

be the fact, in the late publication of his ingenious Essays on Natural History *.

Woodcocks likewise are known to build in some parts of England every year; but, as the instances are commonly those of a single nest, I would by no means pretend to draw the same proof against the summer migration of this bird, as in the former case of the snipe.

I will most readily admit, that these accidental facts are rather to be accounted for, perhaps, from the whimsy or silliness of a few birds, which occasions their laying their eggs in a place where they are easily discovered, and contrary to what is usual with the bulk of the species.

I remember to have seen a duck's nest once on the top of a pollard willow, near the decoy in St. James's Park; it would not be, however, fair to infer from such an instance, that all ducks would pitch upon the same very improper situation for a nest, upon which it is difficult to conceive how a web-footed bird could settle.

Some silly birds likewise now and then choose a place for building, which cannot escape the observation of either man or beast, as he passes by.

I therefore suppose that the few proofs of woodcocks nests having been found in England, arise either from one or other of these two causes, and all which they seem to prove is, that our climate in summer is not absolutely improper for them.

It is to be observed, however, that Mr. Catesby considers such instances as of equal force against the

* P. 72.

migration

migration of the woodcock, as of the snipe *. Willughby also says, that Mr. Jessop saw young woodcocks sold at Sheffield (which rather implies a certain number being brought to market), and that others had observed the same elsewhere †.

We are, indeed, informed by Scopoli ‡, that they breed constantly in Carniola, which is considerably to the southward of any part of England : our country is therefore certainly not too hot for them.

Woodcocks appear and disappear almost exactly about the same time in every part of Europe, and perhaps Africa ‖ : heat and cold, therefore, seem not to have any operation whatsoever with regard to the supposed migration of this bird.

But it may be said, what signifies proving the probability of woodcocks breeding in England, if it is not a known fact that they do so ?

To this it should seem there are several answers, as it is equally incumbent upon those who contend for migration, to shew that these birds were ever seen on such passage.

Another answer is, ask ninety-nine people out of a hundred, whether snipes ever make a nest in Enggland; and they will immediately say, that they do not ; so little are facts or observations of this sort attended to.

But I shall now endeavour to give some other reasons why woodcocks may not only continue with us

* Phil. Tranf. abr. Vol. II. p. 889.
† B. iii. c. 1.
‡ Ornith. Leipfig, 1769.
‖ Shaw's Trav. Phyf. Obf. ch. ii.

during the fummer, but alfo breed in large tracts of wood or bog, without being obferved.

In the other parts of Europe, all birds almoft are confidered as game, or, at leaft, are eaten as wholefome food, Ray therefore mentions, that hawks and owls are fold by the poulterers at Rome; every fort of fmall bird alfo is equally the foreign fowler's object *.

An Englifhman does not confider, on the other hand, perhaps twelve kinds of birds worthy his attention, or expence of powder, none of which are ever fhot in our woods during the fummer, nor are birds then difturbed by felling either coppice or timber.

But it will be faid, why are not woodcocks fometimes feen, however, as they may be fuppofed to leave their cover in fearch of food ?

To this I anfwer, that woodcocks fleep always in the daytime, whilft with us in the winter, and feed only during the night †. Whenever a woodcock, therefore, is flufhed, he is roufed from his fleep by the fpaniel or fportfman, and then takes wing, becaufe there are no leaves on the trees to conceal the bird.

Whoever hath looked attentively at a woodcock's eye, muft fee that, from the appearance of it, the

---

* In one of Boccace's Novels, a lover, who lives at Florence, dreffes a falcon for the dinner of his miftrefs. Giorna a V. Novel. IX.

† Almoft all the wild fowl of the duck kind alfo fleep in the daytime, and feed at night.

sight must be more calculated to distinguish objects by night than by day *.

The fact therefore is notorious to those who cut glades in their woods, and fix nets for catching these birds, that they never stir but as it begins to be dark, after which they return again by day-break, when their sight even then is so indifferent, that they strike against the net, and thus become entangled.

No one with us ever thinks of fixing or attending such nets in summer for woodcocks, because it is not then supposed that there is any such bird in the island; if they tried this experiment, however, I must own that I believe they would have sport †.

Mr. Reinhold Forster, F. R. S. who is an able naturalist, informs me, that the fowlers in the neighbourhood of Dantzick kill many woodcocks about St. John's day (or Midsummer), in the following man-

* I conceive also, it is from the eyes looking so dull; that this bird is generally considered as being so foolish : hence the Africans call the woodcock *hammar el hadgel*, or the partridge's ass. Shaw's Phyf. Obf. ch. ii.

† I would ask those who will probably laugh at the very idea of such sport (which I do not, however, absolutely insure), whether, if I was to send them to any part of the British coast to catch the true anchovy, or tunny fish, they would not suppose equally that it was a fool's errand.

Notwithstanding, however, this incredulity, I can produce the authority of both Ray (Syn. Pifc. p. 107.) and Mr. Pennant (Brit. Zool. ill. p. 34. 36.), that the true anchovy is caught in the sea not far from Chester, and the tunny fish on the coast of Argyleshire, together with the herrings, where they are called *mackrel fture*.

Is it not amazing, however, that a fish of such a size as the tunny should never have been heard of, even by the Scotch naturalist Sir Robert Sibbald ?

ner,

ner, and that they continue to do so till the month of August.

They wait on the side of some of the extensive woods in that neighbourhood, before day-break, for the return of the woodcock from his feeding in the night-time, and always depend upon having a very good chance of thus shooting many of them.

The Dantzickers, however, might be employed the whole summer near these woods in the day-time, without ever seeing such a bird; and it seems therefore not improbable, that it arises from our not waiting for them at twilight or day-break, that they are never observed by Englishmen in the summer. If this bird should, however, be seen in the night, it is immediately supposed to be an owl, which a woodcock does not differ much from in its flight.

To these reasons for woodcocks not being observed, it may be added, that the bird is believed to be absolutely mute, and consequently, never discovers itself by its call.

If it be still contended, that the nest or young must sometimes be stumbled upon, though in the centre of extensive woods, or large bogs, the siskin (or aberdavine *) is a much more extraordinary instance of concealing its nest and young.

The plumage of this bird is rather bright than otherwise; and the song, though not very pleasing, yet is very audible, both which circumstances should discover it at all times; yet Kramer † informs us, that, though immense numbers breed annually on

* Brit. Zool. p. 309.
† Elenchus Animalium per Austriam, p. 261. Viennæ, 1756.

I   the

the banks of the Danube, no one ever obferved the
neft.

This bird is rather uncommon in England; fo that
if 1 afk when the neft was ever found within the
verge of the ifland, it may be confidered as rather
an unfair challenge.

There is another bird, however, called a red-
poll *, which is taken in numbers during the Michael-
mas and March flights by the London bird-catchers,
whofe neft, I believe, was never difcovered in England, though I have feen them in pairs during the
fummer, both in the mountainous parts of Wales
and highlands of Scotland †.

But I fhall now mention another proof that woodcocks breed in England.

The Reverend Mr. White, of Selborn, who is
not only a well-read naturalift, but an active fportfman, informs me, that he hath frequently killed
woodcocks in March, which, upon being opened,
had the rudiments of eggs in them, and that it is
ufual at that time to flufh them in pairs. Willughby
alfo obferves the fame ‡.

This bird, therefore, certainly pairs before its
fuppofed migration; and can it be corceived that
this ftrict union (which birds in a wild ftate fo faith-
fully adhere to) ‖, fhould take place before they

* Brit. Zool. p. 312.

† This e'egant little bird is very common in Hudfon's Bay,
where it feeds chiefly on the birch trees; which being more
common in the northern than fouthern parts of Great Britain,
may account for the bird's being more often feen northward.

‡ B. III. c. i.

‖ It is believed that no mule-bird was ever feen in a wild
ftate, notwithftanding M. de Buffon fufpects many an intrigue

traverfe

traverfe oceans, and when they cannot as yet have pitched upon a proper place for concealing their neft and neftlings?

Let us examine if this intercourfe before migration takes place in other birds, which are fuppofed to crofs wide extents of fea : and a quail affords fuch proof.

I have been prefent when thefe birds have been caught in the fpring, which always turn out to be males, and are enticed to the nets by the call of the hen; quails therefore pair after they appear in England.

But I fhall now confider the other two inftances of birds which are feen with us in the winter, and are not obferved in the fummer; I mean, the fieldfare and redwing.

And firft, let us examine, where thefe birds are actually known to breed : the northern naturalifts fay, in Sweden; Klein, in the neighbourhood of Dantzick, which is only in lat. 54° 30′ *; and Willughby, in Bohemia.

in the receffes of the woods (Hift. Nat. des Oifeaux, tom. I.) fuch irregular intercourfe is only obferved in cages and aviaries, where birds are not only confined, but pampered with food.

* See Klein, de Avibus Erraticis, p. 178. Klein, however, cites Zornius, who lived in the fame part of Germany, and who afferts that the *turdus Iliacus* (or redwing) leaves thofe parts in the fpring. The circumftance therefore of the redwing's breeding in numbers *(per multitudines)* had efcaped the notice of Zornius, though he hath written a differtation on this queftion.

Is it at all furprizing, after this, that fuch difcoveries, if made at all, fhould not be commonly heard of?

As they therefore build their nests in more Southern parts of Europe, there is certainly no natural impoffibility of their doing fo with us, though, I muft own, I never yet heard but of one inftance, which was a fieldfare's neft found near Paddington *.

I cannot, however, but think it is only from want of obfervation, that more of fuch nefts have not been difcovered, which are only looked after by very young children; and the chief object is the eggs, or neftlings, not the bird which lays them †.

The plumage therefore and flight of the fieldfare or redwing being neither of them very remarkable, it is not at all improbable they may remain in fummer, without being attended to; and particularly the redwing, which fcarcely differs at all in appearance from other thrufhes. Thus the cough is by no means peculiar to Cornwall, as is commonly fuppofed, but is miftaken for the jackdaw, or rook.

But it may be faid, that thefe birds fly in flocks during the winter, and if they remain here during the fummer, we fhould fee them equally congregate.

I have not before referred to Klein, who hath written a very able treatife, in which he argues againft the poffibility of migration in birds; becaufe, though I fhould be very happy to fupport my poor opinion by his authority, yet I thought it right neither to repeat his facts, or arguments.

* See alfo Harl. Mifc. Vol. II. p. 561.

† Many birds alfo build in places of fuch difficult accefs, that boys cannot climb to; birds-nefting is confined almoft entirely to hedges, and low fhrubs.

This

This circumstance, however, is by no means peculiar to the fieldfare and redwing; moſt of the hard-billed ſinging birds do the ſame in winter, but ſeparate in ſummer, as it is indeed neceſſary all birds ſhould during the time of breeding.

I ſhall now conſider another argument in favour of migration, which I do not know hath been ever inſiſted upon by thoſe writers who have contended for it, and which at firſt appearance ſeems to carry great weight with it.

There are certain birds, which are ſuppoſed to viſit this iſland only at diſtant intervals of years; the Bohemian chatterer and croſs-bill * (for example) once perhaps in twenty.

The fact is not diſputed, that ſuch birds are not commonly obſerved in particular ſpots from year to year ; but this may ariſe from two cauſes, either a partial migration within the verge of our iſland, or perhaps more frequently from want of a ready knowledge of birds on the wing, when they happen to be ſeen indeed, but cannot be examined.

I never have diſputed ſuch a partial migration; and indeed I have received a moſt irrefragable proof of ſuch a flitting, from the Rev. Mr. White of Selborn in Hampſhire, whoſe accurate obſervations I have before had occaſion to argue from.

---

* This bird changes the colour of its plumage at different ſeaſons of the year, which is ſometimes red.

The firſt account we have of their being ſeen, is in the Ph. Tr. abr. Vol. V. p. 33. where Mr. Edward Lhwyd ſuſpects them to be Virginia nightingales, from their feathers being red, and had no difficulty of at once ſnppoſing that they had croſſed the Atlantic.

The

The rock (or ring-ouzel) hath always hitherto been confidered as frequenting only the more mountainous parts of this ifland: Mr. White, however, informs me that there is a regular migration of thefe birds, which flock in numbers, and regularly vifit the neighbourhood of Selborn, in Hampfhire *.

I therefore have little doubt but that they equally appear in others of our Southern counties; though it efcapes common obfervation, as they bear a fort of general refemblance to the black-bird, at leaft to the hen of that fpecies.

I own alfo, that I always conceived the Bohemian chatterer was not obferved in Great Britain but at very diftant intervals of years, and then perhaps only a fingle bird, whereas Dr. Ramfey (profeffor of natural hiftory at Edinburgh) informs Mr. Pennant, that flocks of thefe birds appear conftantly every year in the neighbourhood of that city †.

As for crofs-bills, they are feen more and more in different parts of England, fince there have been fo many plantations of firs: this bird is remarkably fond of the feeds of thefe trees, and therefore changes its place to thofe parts where it can procure the greateft plenty of fuch food ‡.

---

* See alfo Br. Zool. Ill. p. 56.

† Thefe birds are faid to be particularly fond of the berries of the mountain-afh, which is an uncommon tree in the Southern parts of Great Britain, but by no means fo in the North.

‡ This bird fhould alfo, for the fame reafon, be found from year to year in the cyder counties, if it was true (as is commonly fuppofed) that he is particularly fond of the kernels of

This

This flitting therefore by no means amounts to a total and periodical migration over feas, but is no more than what is experienced with regard to feveral birds.

For example, the Britifh Zoology informs us *, that, at an average, 4000 dozen of larks are fent up from the neighbourhood of Dunftable, to fupply the London markets ; nor do I hear, upon inquiry, that there is any complaint of the numbers decreaf- ing from year to year, notwithftanding this great confumption.

I fhould not fuppofe that 50 dozen of fkylarks are caught in any other county of England ; and it fhould therefore feem that the larks from the more adjacent parts croud in to fupply the vacuum occafion- ed by the London Epicures, which may be the caufe poffibly of a partial migration throughout the whole ifland.

I begin now to approach to fomething like a con- clufion of this (I fear) tedious differtation : I think, however, that I fhould not omit what appears to me at leaft as a demonftration, that one bird, which is commonly fuppofed to migrate acrofs feas, cannot poffibly do fo.

apples, which it is conceived he can inftantly extract with his very fingular bill.

Mr. Tunftall, F. R. S. however, at my defire, once placed an apple in the cage of a crofs-bill, which he had kept for fome time in his very valuable and capital collection of live birds : upon examining the apple a fortnight afterwards, it remained untouched.

* P. 235.

A landrail

A landrail *, when put up by the shooter, never flies 100 yards; its motion is excessively slow, whilst the legs hang down like those of the water fowls which have not web feet, and which are known never to take longer flights.

This bird is not very common with us in England, but is excessively so in Ireland, where they are called corn-creaks.

Now those who contend that the landrail, because it happens to disappear in winter, must migrate across oceans, are reduced to the following dilemma.

They must first either suppose that it reaches Ireland periodically from America; which is impossible, not only because the passage of the Atlantic includes so many degrees of longitude, but because there is no such bird in that part of the globe.

If the landrail therefore migrates from the continent of Europe to Ireland, which it must otherwise do, the necessary consequence is, that many must pass over England in their way Westward to Ireland; and why do not more of these birds continue with us, but, on the contrary, immediately proceed across the St. George's channel?

Whence should it arise also, if they pass over this island periodically in the spring and autumn, that they are never observed in such passage, as I have already stated their rate in flying to be excessively slow; to which I may add, that I never saw them rise to the height of twenty yards from the ground, nor indeed exceed the pitch of a quail.

* Br. Zool. p. 387.

I have

I have now fubmitted the beft anfwers that have occurred, not only to the general arguments for the migration of birds acrofs oceans, but alfo to the particular facts, which are relied upon as actual proofs of fuch a regular and periodical paffage.

Though I may be poffibly miftaken in many of the conjectures I have made, yet I think I cannot be confuted but by new facts, and to fuch frefh evidence, properly authenticated, I fhall moft readily give up every point, which I have from prefent conviction been contending for.

I may then perhaps alfo flatter myfelf, that the having expreffed my doubts with regard to the proofs hitherto relied upon, in fupport of migration, may have contributed to fuch new, and more accurate obfervations.

It is to be wifhed, however, that thefe more convincing and decifive facts may be received from iflanders (the more diftant from any land the better *) and not from the inhabitants of a continent; as it does not feem to be a fair inference, becaufe certain birds leave certain fpots at particular times, that they therefore migrate acrofs a wide extent of fea.

For example, ftorks difappear in Holland during the winter, and they have not a very wide tract of fea between them and England; yet this bird never frequents our coafts.

---

* I would particularly propofe the iflands of Madera and St. Helena; to thefe, I would alfo add the ifland of Afcenfion (had it any inhabitants), as likewife Juan Fernandez, for the Pacifick ocean.

The

The ftork, however, may be truely confidered as a bird of paffage, by the inhabitants of thofe parts of Europe (wherever fituated) to which it may be fuppofed to refort during the winter, and where it is not feen during the fummer.

I am, dear Sir,

Your moft faithful,

humble fervant,

Daines Barrington.

P. S.

P. S.

SINCE I fent to you my very long letter on the migration of birds, I have had an opportunity of examining the " Planches Enluminées," which are faid to be publifhed under M. de Buffon's infpection, and which feem to afford a demonftration of M. Adanfon's inaccuracy in fuppofing either the roller, or fwallows, which he caught in his fhip, near the coaft of Senegal, to be the fame with thofe of Europe.

In the 8th of thefe plates, there is a coloured figure of a bird, called le rollier d'Angola, which agrees exactly with M. Adanfon's defcription * ; but he trufted too much to his memory, when he pronounced it to be the fame with the Garrulus Argentoratenfis of Willughby, and therefore fuppofed it to be on its paffage to Europe.

This bird hath, indeed, in many refpects, a very ftrong refemblance to the common roller of Europe, which is reprefented alfo in the Planches Enluminées, plate 486; but it differs moft materially in the length of the two exterior feathers of the tail, as well as in the colour of the neck, which in the African roller is of a moft bright green, and in the European of rather a dull blue.

In the 310th plate, there is likewife a coloured reprefentation of the " Hirondelle a ventre roux du " Senegal," which fpecimen was poffibly furnifhed by Monf. Adanfon himfelf.

* Voyage au Senegal, p. 15. There is alfo another African bird, reprefented in the " Planches Enluminées," which might very eafily, on a hafty infpection, be miftaken for the Garrulus Argentoratenfis, viz. the Guepier a longue queue du Senegal. Pl. Enl. p. 314.

The roller of Angola is alfo engraved by Briffon, T. ii. pl. 7.

It very much refembles the European fwallow, but the tail differs, as the forks (in the Senegal fpecimen) taper from the top of the two exterior feathers to the bottom, at three regular divifions, whereas in the European they are nearly of the fame width throughout.

The convincing proof, however, that the " Hi- " rondelle a ventre roux du Senegal" differs from our chimney fwallow is, that the rump is entirely covered with a bright orange or chefnut, which in the European fwallow " is of a very lovely but dark " purplifh blue colour *."

Having lately looked into Ariftotle's Natural Hiftory, with regard to the cuckow, I take this opportunity alfo of enlarging on the doubts I have thrown out, in relation to the prevailing notion of this bird's neftlings being hatched and fed by fofter parents.

I find that this moft general opinion takes its rife from what is faid by this father of natural hiftory, in his ninth book, and twenty-ninth chapter.

Ariftotle there afferts, that the cuckow does not build a neft itfelf, but makes ufe moft commonly of thofe of the wood-pigeon, hedge-fparrow, lark, (which he adds are on the ground) as well as that of the χλωρις †, which is in trees.

Now, if we take the whole of this account together, it is certainly not to be depended upon; for the wood-pigeon ‡ and hedge-fparrow do not build upon the ground, and no one ever pretended to have

---

* See Willughby, p. 312.

† The χλωρις is rendered *luteola*; but, as there is no defcription, it is difficult to fay what bird Ariftotle here alludes to; Zinanni fuppofes it to be the greenfinch.

‡ The wood-pigeon, from its fize, feems to be the only bird which is capable of hatching, or feeding, the young cuc-

found a cuckow's egg in the neſt of a lark, which, indeed, is ſo placed.

I have before obſerved, that the witneſſes often vary with regard to the bird in which the cuckow's egg is depoſited *; and Ariſtotle himſelf, in the ſeventh chapter of his ſixth book, confines the foſterparents to the wood-pigeon and hedge-ſparrow, but chiefly the former.

If the age † of Ariſtotle is conſidered, when he began to collect the materials for his Natural Hiſtory, by the encouragement of Alexander after his conqueſts in India ‡, it is highly improbable he ſhould have written from his own obſervations. He therefore ſeems to have haſtily put down the accounts of the perſons who brought him the different ſpecimens from moſt parts of the then known world.

Inaccurate, however, and contradictory as theſe reports often turn out, it was the beſt compilation which the ancients could have recourſe to; and Pliny

kow; yet, if it is recollected that this bird lives on feeds, it is probable that the cuckow, whoſe nouriſhment is inſects, would either be ſoon ſtarved, or incapable of digeſting what was brought by the foſter-parent. This objection is equally applicable to the χλωρις, if it is our greenfinch.

* Thus Linnæus ſuppoſes it (in the Fauna Suecica) to be the white wagtail, which bird builds in the banks of rivers, or roofs of houſes, (See Zinanni, p. 51.) where it is believed no young cuckow was ever found.

† He did not leave the ſchool of Plato till the age of thirtyeight (or, as ſome ſay, forty); after which, ſome years paſſed before he became Alexander's preceptor, who was then but fourteen: nor could he have written his Natural Hiſtory, probably, till twelve years after this, as Pliny ſtates that ſpecimens were ſent to him by Alexander, from his conqueſts in India. Ariſtotle therefore muſt have been nearly ſixty, when he began this great work, and conſequently muſt have deſcribed from the obſervations of others.

‡ Pliny, L. viii. c. 16.

there-

therefore professes only to abridge him, in which he often does not do justice to the original.

Whatever was asserted by Ariftotle, is well known to have been moft implicitly believed, till the laft century; and I am convinced that many of the learned in Europe would, before that time, not have credited their own eyefight againft what he had delivered.

There cannot be a ftronger proof that the general notion about the cuckow arifes from what is laid down by Ariftotle, than the chapter which immediately follows, as it relates to the goatfucker, and ftates that this bird fucks the teats of that animal.

From this circumftance, the goatfucker hath obtained a fimilar name in moft languages, though it is believed no one (who thinks at all about matters of this fort) continues to believe that this bird fucks the goat *, any more than the hedgehog does the cow.

I beg leave, however, to explain myfelf, that I give thefe additional reafons only for my doubting with regard to this moft prevailing opinion; becaufe I am truly fenfible that many things happen in nature, which contradict all arguments from analogy, and I am perfuaded, therefore, that the firft perfon who gave an account of the flying fifh, was not credited by any one, though the exiftence of this animal is not now to be difputed.

All that I mean to contend for is, that the inftances of fuch extraordinary peculiarities in animals, fhould be proportionably well attefted, in all the neceffary circumftances.

I muft own, for example, that nothing fhort of the following particulars will thoroughly fatisfy me on this head.

* See Zinanni p. 95. who took great pains to detect this vulgar error.

The

The hedge-fparrow's neft muft be found with the proper eggs in it, which fhould be deftroyed by the cuckow, at the time fhe introduces her fingle egg *.

The neft fhould then be examined at a proper diftance from day to day, during the hedge-fparrow's incubation, as alfo the motions of the fofter parent attended to, particularly in feeding the young cuckow, till it is able to fhift for itfelf.

As I have little doubt that the laft mentioned circumftance will appear decifive to many, without the others which I have required, it may be proper to give my reafons, why I cannot confider it alone, as fufficient.

There is fomething in the cry of a neftling for food, which affects all kinds of birds, almoft as much as that of an infant, for the fame purpofe, excites the compaffion of every human hearer †.

I have taken four young ones from a hen fkylark, and placed in their room five neftling nightingales, as well as five wrens, the greater part of which were reared by the fofter parent.

It can hardly in this experiment be contended, that the fkylark miftook them for her own neftlings, be-

---

* I could alfo wifh that the following experiment was tried. When a hedge-fparrow hath laid all her eggs, a fingle one of any other bird, as large as a cuckow, might be introduced, after which if either the neft was deferted, or the egg too large to be hatched, it would afford a ftrong prefumption againft this prevailing opinion. I muft here alfo take notice, that Mr. Hunter, F. R. S. who hath diffected hen cuckows, informs me that they are not incapacitated from hatching their eggs, as hath been fuppofed by fome ornithologifts.

† I am perfuaded that a cuckow is oftener an orphan, than any other neftling, becaufe, from the curiofity which prevails with regard to this bird, the parents are eternally fhot.

caufe

cause they differed greatly, not only in number and size, but in their habits, for nightingales and wrens perch, which a skylark is almost incapable of, though, by great assiduity, she at last taught herself the proper equilibre of the body.

I have likewise been witness of the following experiment: two robins hatched five young ones in a breeding cage, to which five others were added, and the old birds brought up the whole number, making no distinction between them.

The Aëdologie also mentions (which is a very sensible treatise on the nightingale *) that nestlings of all sorts may be reared in the same manner, by introducing them to a caged bird, which is supplied with the proper food.

Not only old birds, however, attend to this cry of distress from nestlings, but young ones also which are able to shift for themselves.

I have seen a chicken, not above two months old, take as much care of younger chickens, as the parent would have shewn to them which they had lost, not only by scratching to procure them food, but by covering them with her wings; and I have little doubt but that she would have done the same by young ducks.

I have likewise been witness of nestling thrushes of a later brood, being fed by a young bird which was hatched earlier, and which indeed rather over-crammed the orphans intrusted to her care; if the bird however erred in judgement, she was certainly not deficient in tenderness, which I am persuaded she would have equally extended to a nestling cuckow.

* Paris, 1751, or 1771.

XXII. ΚΟΣ-